Pig Showing

-

Containing Information on Judging, Preparation and Handling Pigs for Exhibition

By

Various Authors

Contents

ADVANTAGES OF SHOWING

For the breeder of pure-bred hogs who wishes to establish a substantial trade, it is very necessary that, as soon as he becomes well versed and properly started in his breeding line, he should make it a practice each year to fit a show herd. Possibly at first it is as well for him to prepare only for county shows, until he thoroughly learns what is required to win at the big shows and expositions. There is no way, in my opinion, in which a young breeder can so quickly get before the people and started to selling his hogs for breeders as to annually fit enough animals to fill the classes usually provided for in the various premium lists.

The reason I suggest that the beginner commence his show career at the county fairs, is from the fact that I passed through all these little troubles when I was a great many years younger than I am now and know what the difficulties are for a beginner. It is hardly expected that a young breeder who has never followed the practice of showing, and who has probably not spent a very large amount of money in the foundation of his herd, can win at state fairs and expositions where only a few great show animals can get the money. Let the county fairs be the stepping stones to larger ones where it requires, practically speaking, the P. T. Barnums of the business to win.

Never will I forget the time way back in the '70's when I made my first show at a state fair. It was at the Illinois State Fair when it went around on wheels, and that year was located at Freeport. I had only been in the business a year; knew nothing of what it required to even have a chance of getting into the money; but nevertheless I was full of enthusiasm and overflowing with ignorance. I fitted up a portion of a herd which I thought was "some pigs" but found, much to my profit eventually, that they were only ordinary. Starting out with much pride and having already figured the amount of money I would bring home by adding up the amounts in the premium lists, I found after the fair was over, that I was really a wiser man and richer in knowledge, but poorer financially than when I arrived on the grounds, beaming with confidence. I did not even get in sight of the premium money. Those were the days when a young breeder was hardly noticed, but, being made of the kind of stuff that never gives up, but sticks, I did not parade the grounds condemning the judges for lack of knowledge, nor inform the managers of the state fair that I would never show again at their fair; but quietly studied the conformation of the animals that were good enough to win, then returned home with the determination to come a little stronger next year.

This was followed up several years before getting much above the white and yellow ribbons. The only blue thing I found in those days was my feelings in not being able to win. This is where I made a mistake by attempting, ignorant as I was, to show at the great state fairs, rather than starting at county fairs. But the determination was in me and the show ring was followed, as large as it was and as great as the shows were, from 1877 until 1893, when the climax was reached at the Chicago World's Fair. As I grew in knowledge and experience each year, I was able to win a few of the better prizes and as the years came and went, won more prizes, until it was a pretty sure thing when I started out on an eight weeks' trip to the big state fairs, that I would win money enough to pay all expenses and more and be benefited greatly by building a substantial acquaintance among breeders in our line.

The trade grew annually and after winning the grand championship at the World's Fair in Chicago for the best herd, consisting of one boar and four sows, over one year old, my name was finally placed on the map and my son and I have practically discontinued showing since that time.

I strongly urge the show ring as a means, not only of education for the breeder, but of building up a substantial business.

SELECTING AND FITTING THE SHOW HERD

The selection of the show herd requires a knowledge of what it takes, when well fitted, to win. One should commence months in advance of the show to make his selections, first making up his mind whether or not he wishes to fill all the various classes listed in the premium lists, which are now so made up that the same animals can show throughout the circuit without being required to carry along other animals of various ages from those first selected to fill the classes.

In making the selection for the show herd, commence first by selecting the animals that are required to show in the "aged class."

The aged sow class should be made up of animals that have proven themselves breeders and should show by their appearance that they have been breeders. Let them, however, be well fitted without overdoing and as uniform as possible in type and conformation, with an aged boar of the same general type, showing that he has been a breeder, only of a more masculine appearance, thus making up a desirable herd for the aged class.

The tusks of the boar should be removed before starting out on the show circuit and should be so closely cut that no trace of the same can be seen. This should be done both for the safety of the caretaker and those about the show ring.

Next, select the senior yearlings, a boar and three sows—which should carry as much size, or nearly so as the aged herd, but would naturally be somewhat smoother owing to the difference in age and previous service. Be sure that these are also uniform in type with the aged herd.

Next select the junior yearling—a boar and three sows, which are, generally speaking, the sweetest things in the show ring, and being at an age that shows full development and yet not required to have produced any offspring, naturally will be much smoother and more in bloom. This class is usually the best of all at any breed show. These should be of the same general type as the older ones selected, and should be in the pink of show condition—well developed at every point—carrying all the flesh that goes to make an ideal show animal, yet under no circumstances to be overly fatted or fleshed to a condition of unevenness or roughness.

This same careful selection must be carried out also in the boar and three sows under twelve months of age. This is really the hardest class to fill satisfactorily. First, for the reason that the young boar over six and under twelve months of age is passing through a crisis in his development; this being an age that almost invariably, if the boar is a vigorous one, finds him fretting and champing more or less when in sight of other animals. The sows, also of this age, are harder to properly develop and bring out in their best form, as they too are passing through a period of life when there is more or less restlessness and excitement than when older.

Some people greatly enjoy bringing out a pig herd, under six months of age, as there are often enticing prizes offered in this class, and it is a good way to show the class of pigs you are producing. While this class should also be as uniform in type and conformation as the others, and should be well fitted in flesh, care should be taken that they are not pushed too hard and become overdone, and more or less wrinkled in appearance.

Fitting the Show Herd.—We will now suppose that the herd has been selected along proper lines and we are ready to start the fitting process with a bunch of animals of the same type. The question now is how to fit these animals to the best advantage without overdoing them, so they will show when the fair season opens, in the best possible bloom.

Always remember that "bloom" is a desirable thing in a show animal. When "in bloom" a hog is at its best and this "bloom" only lasts for a short time, and is as easily lost as the bloom of a ripe peach, hence the matter of fitting should commence in time and be carried on in a manner to have the animals "in bloom" when the fair season opens.

For convenience in handling and caring for them, the show animals should be kept by themselves and not allowed to roam with the general herd. The four boars being fitted for the different herds should be carefully prepared and put together in one lot where they can be fed and handled together so that there will be no danger of their fighting should it be necessary to drive them to or from show building together. The aged sows, the senior and junior yearlings, should also be fed and kept in one enclosure for best results during the fair circuit, when it is necessary to exercise them mornings during the show season. The over six months and under six months sows can also be kept together for the same reasons.

The yards or lots, where the animals are to be fitted during the next few months, should have an abundance of succulent pasture, either natural grasses or other green forage crop provided and each lot should have a water fountain that should be kept well supplied with pure clean water at all times. Of course ample shade should be provided—either natural or artificial—and no dust should be allowed to accumulate where these animals lie in the shade. If possible, a cement bathing pool large enough and deep enough in the deepest part to practically cover them should be provided, in which the water should be changed every few days, always remembering to add a little disinfectant and crude oil to the water, preventing any skin troubles or lice.

Feeding the Herd.—When one starts out to fit a herd of show animals for the large fair circuits, the matter of expense must, in a measure be forgotten as these animals should have the best feed possible to put them in just the right condition to show all that is in them. We have found, when fitting animals for show, that we have to make some changes in our regular grain mixtures for best results. I have found nothing better in this case than to feed a nice rich slop feed, made up about as follows:

For the older animals above one year I would use equal parts of finely ground corn meal and the best quality of white heavy middlings, with an addition of ten per cent old process oil meal, thoroughly mixed together before being wet, and if sweet skimmed milk or that from the separator is possible to be had at any reasonable price, the feed should be mixed with this and all feed in a sweet condition. If it is found impossible to get milk, add tankage to the oil meal (eight to ten percent oil meal and five percent tankage), and the mixture of meal and middlings, and mix this with fresh water and feed it after it soaks a few hours—but never allow it to become sour.

Two Feeds a Day.—I think two feeds a day, even in the fitting of a show herd, is as good as three, generally; however, it would be proper and perhaps as well to give these animals a little soaked shelled corn that has soaked long enough to become soft, as a noon-day meal. Where this is done, the morning and evening feed can be fed a little earlier and later, respectively.

For the younger herds of over six and under twelve months, as well as the herd under six months, I should certainly urge the use of skimmed sweet milk for best results. I have known showmen, while fitting young pigs, to feed whole milk fresh from the cow. This is entirely useless and, I think, is one of the surest ways of overdoing the pig and causing him to break down in the pasterns and it is also almost a certainty that pigs fed fresh warm whole milk will become more or less wrinkled, but skim milk is not so apt to cause this trouble, and this is especially true when well balanced with the ground feed.

Always remember, however, that too much milk is worse than none. The proper amount is three pounds milk to one pound of grain; with this your ration is practically balanced or at least gives the best results.

Condiments.—If any of the animals at any time during the fitting do not eat with as much relish as would seem best, there are many kinds of condiments that could be used to make the feed more palatable. A sufficient amount of brown sugar or black strap molasses to well sweeten the mixture, will make the feed much more palatable. This would be the case even when milk was used instead of water in mixing the feed.

I would, by all means urge, while fitting these hogs for show, that a mixture of mineral matter (mentioned in another part of this book) be used. It may be well to mention here that this should be made up of ground limestone, with perhaps some slacked lime, together with ground phosphate rock or any other material that would contain plenty of phosphate and the whole mixed with salt to make it more palatable. I urge the use of this material so that while fleshing these animals there would be no danger of breaking down the bone, as the mineral matter is essential for bone growth.

If any of the younger animals in the show herd were inclined to be a little weak in the feet or pasterns, I would buy one hundred pounds of finely ground bone meal or bone dust and mix it with the mineral matter or even add it in small quantities to the slop feed.

Exercise Necessary.—Much care should be taken in fitting the show herd not to break them down, and as a help in this matter as well as in keeping them in better condition, it is necessary that the show herd be given some special exercise other than that which they will naturally take in their enclosure. By taking the three older boars out each day and driving a half mile or more you will not only have them under perfect control, but will also have them in a condition to show at their best when driven in the show ring, and the same is true of the show sows and even the under a year herds.

We have all noticed, while standing about a show ring and watching the animals come in from their pens, that many of them, while in good flesh, are not really in show condition, for the reason that they cannot walk with ease and grace but wabble around. A show animal should not be so heavily loaded with fat as not to be able to walk with ease and comfort. It is not always the amount of flesh that an animal is carrying that makes it show at its best, but the smoothness in which it is put on and the ease with which it is carried.

It is my opinion that what is known as "strong breeding condition", that is, flesh enough to round out every point without overloading, is the proper show condition.

What I have said about the feed to be used in fitting the show herd, need not be considered as an absolute iron-clad rule. Any of the mixtures of grain, grasses, etc., that will come close to being a balanced ration is all that is necessary. I merely name these feeds as among what we have found to be the most satisfactory. The real object is to feed what will flesh them rapidly and not add too much fat, but more red meat or flesh.

Finishing Touches.—While fitting the herd for the show ring they should be handled daily by the feeder, by brushing them a little or scratching them and coaxing them to lie down where he can handle them about the legs and feet, so that when you are ready to trim the toes and hoofs into nice shape they will not get excited, but will lie quietly and let you work over them as you wish.

The foot and pastern of the show pig can be improved one hundred percent by proper trimming. When the pig is lying down, quietly take the foot in the left hand and with a very sharp knife trim the lower edges off the hoof, commencing well back and following around the entire hoof, shaping the toe up as close as possible to the fleshy part of the foot without injury. If the dewclaws are of unusual length these too can be shaped up at the bottom and pared down to proper shape. All this work should be done at odd times before starting out to the fair.

A nice brushing every day or two after sprinkling with disinfectant and crude oil is very essential, not only to make the show herd quiet and docile, but to improve the condition of the skin and hair. All this is a help in shedding the old coat. The earlier this is commenced in the preparation of fitting a show herd the better. Every animal except the under six months pigs should shed off his or her old coat not later than the first of August, that the new coat may be nicely started before the fair season opens. Generally all this will come along in due time if the animals are fed as above and are gaining in flesh constantly. If any of the herd should not begin to show inclination to shed by the middle of June or the first of July, I would give them a wallow hole in which some clay has been placed, if it is not naturally a clay soil, keeping this hole rather thick in mud, and adding some wood ashes.

Clipping the Hair.—Many showmen are in the habit of clipping the hair of their older animals when they do not shed off in time. While this, in some instances, looks better than an extremely coarse coat of hair, it always shows every little unevenness in the flesh of the animal. This practice is more common among the Poland-China showmen than any other breed I believe, yet I have seen some show animals come into the ring that were closely clipped, showing almost no hair and sprayed in oil, that really I think were not showing as well as though not clipped, for the reason that little uneven places could be plainly seen along the back and sides, evidence to the Judge that they did not flesh evenly as they should, and would in a way, militate against them.

Before entering the show ring or as early after arriving on the fair grounds as possible, the herdsman should take a hand clipper and clip the long hairs off the edges of the ears and about the nose and jaw of the hogs and also clip the tail clean from the brush back to the tail head, giving a much more finished appearance to the animal than though this was neglected. The above suggestions properly followed and the bringing of the herd to the shows in a thoroughly docile, well mannered condition, add much to their credit while in the show ring. It is pretty hard for a Judge to properly examine an aged boar or one even younger, if he is brought into the ring with four or five men, each bumping him around with a short hurdle—the boar certainly is not showing to the best advantage.

Dressing.—A nice dressing to use after the hogs are fitted and in show condition, before entering the ring, is made as follows:

Take a good quality of cotton seed oil, adding enough wood alcohol to thoroughly cut and make a nice thin easy running dressing. After the hog is thoroughly washed and his skin is clean apply with a brush and rub it in thoroughly.

One of the most detestable dressings that I have ever come in contact with as judge at the great shows is made of oil and lamp black. The animals, as they come into the show ring, are not only a mass of grease and lamp black, but the attendants are about as badly blacked up as the hogs, and before the judge is half through he is also more or less greased up. I have known of cases where the judge had to send his clothes to be cleaned each night or put on a clean pair of overalls each day. All that is necessary as a dressing is something that will make the hair glossy and yet not be gummy.

Exercise on the Show Circuit.—The good herdsman and caretaker does not lie in bed until late in the morning, but is up early and has his show animals out on the grass somewhere about the fair grounds, and drives them around for an hour until each animal is thoroughly emptied out and has had proper exercise.

HANDLING SHOW HERD IN THE RING

The proper fitting and handling of the show herd before it starts out on the circuit, will prevent much trouble in handling the animals in the ring.

With the herd properly trained, there is nothing with which to handle them compared with a buggy whip, in the hands of a man who has sense enough not to whip the hogs, but quietly touch them on either side of the head to place them where he wishes. As a matter of fact this has been my experience in the many years of handling show hogs. I never need a hurdle with our hogs. With some breeds it is absolutely necessary to have a hurdle in handling a mature boar even though he is supposed to be well mannered and docile, but there is no excuse in using a hurdle with a bunch of sows if they are half way prepared before starting on the circuit. When a hurdle must be used, let it be a light one and made so that the hog cannot see through it. Don't make it of narrow slats a few inches apart, but cover it with heavy material, or else make it of boards tightly matched so there can be no seeing through it. When in the ring with the herd or a single animal, show to the best possible advantage. The showman has this privilege.

Feeding on Circuit.—Many exhibitors seem to think that when they start out on the show circuit they must stuff the animals with all the feed possible, not only during the time they are on the cars going to and from the shows but each day while on the grounds. It has been our experience that the man who follows this custom generally arrives home with his hogs much lighter in weight than when he started out, while if the hogs had been given only water to drink while en route to the shows and fed lightly for the first day after arriving and given plenty of exercise, they would wind up the circuit in much better condition than if they had been stuffed all the time.

I have known an exhibitor to buy warm milk from some of the dairymen and feed his pigs all they could hold, though they had never had a drop at home while being fitted. This generally results in a case of scours with the pigs "off feed" for several days and by the time they go into the ring they are badly gaunted up. Of course if the pigs have had this ration at home it should be continued. Avoid radical changes in the rations.

Treatment of the Herd on Its Return Home.—Many successful exhibitors, when they have finished the show circuit, won their laurels and arrived home safely with their herds, seem to think that the animals now need no further attention, except feed.

This is a great mistake, and if these show animals are expected to go on and prove what they should be, desirable and regular breeders, they must be handled very carefully.

The first thing I would advise on return from the shows, would be to quarantine the show herd on a portion of the farm or some other place where they would not come in contact with the home herd. They should be placed on good, green, succulent pasture, if possible, and if not possible, should have some kind of green feed to take the place of pasture. They should be fed quite a little less than while on the show circuit, and no fat-making feed, and be made to take all the exercise possible, so that they may be reduced in flesh somewhat—not by starving, but by lighter feeding and abundant exercise—and if they have not been too strongly fitted, they will soon be in prime condition to breed.

The show herd should be kept in quarantine about three weeks, and if no symptoms of disease appear by that time, it would be safe to put them with the home herd.

Now that we can procure a reliable hog cholera serum I would advise all hog men making the fair circuits to give each show animal a large dose of serum (no virus) about a week before leaving home for the fairs, unless they have positive knowledge that every animal in their show herd has been properly and permanently immuned by the simultaneous treatment.

A large percent of bran and oats mixed with a small amount of middlings and cornmeal is an excellent feed to use during the reducing period. They must have exercise and if necessary see that they get it by driving daily. This is very important and must not be overlooked.

A part of the ration may consist of whole oats scattered freely in a clean place, as the oats themselves are an excellent feed, and they will get considerable exercise while eating them.

I might say right here that with many exhibitors it is a custom to breed the show sows a month before starting out on the fair circuit, and if successful in settling them, so much the better, even though the litter comes at an unfavorable time of the year. It simply keeps the animals breeding, and it is much better for them.

TO THE EXHIBITOR AND FAIR MANAGER

The wise exhibitor or herdsman will so arrange his circuit that he will arrive on the fair grounds as early before the opening of the fair as possible, that he may have his hogs well rested and in the pink of condition before the show opens. Where one attends a fair each week, this of course is sometimes a hard rule with which to comply, but many thinking fair managers today are so arranging their dates and days of show that the live stock that is to show the following week at a distant state is allowed to be released on Friday night—which, by the way, is a custom that all fairs and expositions should follow.

Many state fairs have too many men among their management who know nothing whatever of the needs of the live-stock exhibitor while on the circuit. They manage their show as though it were the only one the exhibitor was going to attend and seem to think that the exhibitor, because he made an exhibition at their fair, should be obliged to remain there until the last man is gone. They should always remember that without the live-stock exhibit their fair would soon be a thing of the past, and for this reason should give the live-stock exhibitor every encouragement and help possible.

Be Prepared to Show Pedigrees.—Oftentimes in the under-six-months class or the class over six and under twelve months, there is such a wide range of sizes that one hesitates in comparing. Here is where every exhibitor should produce the certificates of registry, and if any of the animals have been purchased of others, the certificates of transfer, showing exact age of the animals on exhibition. This would avoid any unpleasantness between exhibitors or between the exhibitor and the judge.

Authenticating Ages.—It is pretty hard for an experienced judge to step into a ring of pigs showing in the under-six-months class and find most of them of proper size and development, and others showing by their general make-up that they are far beyond six months old, even being old enough to show well developed tusks, which every man knows are not developed until after the pig is six months of age. The judge who knows his business, while not inclined to quarrel with the exhibitor over the age of his pigs, will quietly ignore them, not considering them eligible to the class. This, of course, generally causes the exhibitor to complain when he should be quietly informed that his pigs are out of their class owing to age, and unless he can prove by certificates of registry, properly signed by the record association, he should not be allowed in the ring. For this reason I would urge every prospective showman to always start out fully prepared for such emergencies. Many is the time that I have asked the exhibitor, while acting as judge, the age of his animal; he generally has an answer ready, and when asked if he has his registry papers with him, he replies that he has them at home, but forgot to bring them, and after passing around the ring once or twice, I again ask the gentleman, "What did you tell me the age of this animal is?" and he would give an age entirely different. I have done this on purpose to find out if the man was telling the truth. You know it has been said that it takes an awfully smart man to be a liar.

Again, where registry papers are not absolutely insisted upon, many exhibitors are inclined, when asked the age of under-a-year animals, to give the date of Sept. 1 to 3, as their date of birth, and those in the under-six-months class from March 1 to 3. This of course, has to be taken by the judge as a fact, however much he may doubt or suspect.

This matter of showing pigs of uncertain ages is somewhat in disrepute. It simply puts the man doing business right up against an almost impossible chance of winning, where older pigs than should be admitted to the class are being shown. I know of no way to stop this except by the rigid enforcement of showing certificates of registry.

One may say that the same rascality might be covered up by the owner when sending his pedigree in for registration giving a wrong birth date, showing the animal younger than it really was. When it comes to this proposition the fellow will have to be very smart or he will be tripped up sometime by having registered two litters from the same sow that were born too nearly at the same time.

JUDGING AT SHOWS

There are many good judges; men who not only know the correct type and conformation of show animals of the various breeds, but are men above reproach and can always be relied on to be absolutely square and honest in their decisions. The matter of selecting the best three or five animals, as the rules in the premium list require, is no small task.

The first thing the judge must do when he steps into the arena is to forget all friends and know no man. He must judge the hogs only and let no personal feelings enter his mind. If his brother or son should be showing in the ring he should be a man of strong enough character to turn down their animals, if not worthy, just as quickly as though shown or owned by an entire stranger.

The judge should not attempt to pass on the animal unless he has in his mind a true picture of what the animal of that age and that breed should be. Of course no animal, even a show animal, is perfect.

After carefully examining each animal of the class under view, and finally deciding which, in his opinion, is the best, let him pass this without further attention and consider which is the second best. It is usually much easier to find the first prize animal than the third, fourth or fifth, but after the judge compares points, conditions, general type and conformation, and has his mind made up, let him line these animals up as first, second, third, fourth and fifth as the case may be, for the clerk to take the entry number and write the proper names and the award in the book. Then he should by all means be ready to compare these animals in the presence of the bystanders, explaining why he gave this one the first over that one and so on. It is surprising to a judge sometimes to find what great satisfaction it gives the exhibitor to be shown where his animal lacked in comparison with the one above him, and no judge should act unless he is able to give the reason.

While disliking to speak of myself as a judge, I may be pardoned in saying that I have acted as judge at practically every state fair in the Union and I make it a practice to make this explanation after each decision and many and many a time has the loser come to me and said: "Mr. Lovejoy, I learned more today from what you have shown me about weak points or undesirable ones in hogs than I ever knew before, and I thank you, and I now know I was not entitled to higher honors."

Tricks by Exhibitors.—A judge in a hot ring has many little annoyances; for instance, there is the showman (and it is his right) who brings in an animal that possibly drops a little in the back, with possibly an inclination to sag too much, and while the judge is trying to find this out, the showman is continually bumping the animal on the nose to keep its head down and its back up. It is also amusing often to find an exhibitor who is continually squatting beside or in front of his animal and patting it, trying to attract the attention of the judge to certain points that he thinks might be overlooked. Let me say to this kind of exhibitor that the judge will find all the good points quickly; what the judge is looking for is the weak ones, and if he knows his business he will find them.

Really the best showmen, who are not only good winners but good losers, say very little to the judge unless asked a question, and this is as it should be, for the judge knows that the whole responsibility is on his shoulders and he is willing to take this responsibility without, rather than with, the advice of the owner or showman.

I always like to have the animal that I am judging walk off naturally and without an attendant. I think this quite important, as it will many times show up defects that an expert showman will completely hide from the judge if the showman is allowed to show that animal all the time. With this idea in mind, I invariably stand where I can see the animals come into the ring from their pens.

If you are fortunate enough while showing to win the blue or purple let that joy be confined in your heart; if you are so unfortunate as to lose let that disappointment also be confined in your heart and try to appear a good loser. It really requires a "good sport" to be a good showman, and especially to be a good loser. The judge cannot in his decisions consider the desires or hopes of the exhibitor, but must at all times make the awards according to his judgement and not be influenced by the ringside.

PEDIGREES

This is a matter that does not greatly interest the farmer or feeder who is growing hogs simply for the open market, but must be understood, and thoroughly so, by the breeder of registered hogs who expects to sell a large portion of his produce to other breeders. A pedigree amounts to nothing unless it is a *correct record* of the different blood lines in the sire and dam taken from the established records for the breed. The pedigree in itself adds no value to the individuality of the animal but it is a means of noting the various blood lines that has produced the animal. Neither is a pedigree of any value unless it is made by a man who would under no circumstance write in other than the correct names of animals, with their herd book numbers. A man who would make a false pedigree would do anything else false that came to his mind and should not last long as a breeder of pure-bred hogs of any breed. In other words the pedigree should be a *guarantee* that only such animals were used in producing the particular individual as really were used.

Study Blood Lines.—One who is well versed in the principles of breeding pure-bred animals, and familiar with the value of the different blood lines of the breed, can by studying the pedigree of the animal he purchases to head his herd, know practically what the results will be from using him. The more animals that appear in the pedigree that have made good as producers of superior stock, the better the pedigree and the more valuable it becomes as a guarantee for future quality in the herd; hence it is of great importance that the owner of a high class breeding herd, who wishes to continually improve the quality of his herd, study carefully the pedigree of any new animal that he wishes to introduce as the head of his herd. It is a well known fact, however, that there are few outstanding sires that are worthy of special note in any breed of pure-bred hogs, and buyers of boars, when they order a boar for use, should not expect him to be "one in a thousand," unless he has investigated the animal's get and has proof of the fact that he *is "one* in a thousand," and if such a boar is found he, the buyer, must expect to pay a very large price to secure him, for the owner can ill afford to part with this kind.

Correspondence.—The breeder of pure-bred hogs, after he is established, will have a large amount of daily correspondence to look after. He should make a rule to be prompt in his replies, answering all questions carefully, describing the animal he offers so that the man will not be disappointed, should be order. He should keep a carbon copy of each letter written. He should always keep a letter file of some kind, whereby he can keep each man's correspondence by itself. I think we have every letter received in thirty years, and pasted to it is a carbon copy of the reply. A card system should be kept with the name of each and every correspondent you do business with, and a follow-up card system would be well, as a follow-up letter often results in getting an order that would be overlooked otherwise.

Every breeder should procure a typewriter and learn to use it, and write all letters on this machine. It is rather hard for some men to write a nice hand with a pen, and there are often some words in the letter that are not plainly written and that puzzle the one receiving it; besides, when writing a letter on the machine the copy can be made at the same time and filed with the original letter.

System.—System is a great thing, even in the hog business. Systematic methods of keeping all records, filing all letters, keeping the cards, the breeding records, the feeding records, and everything connected with the business, is most important.

PIG CLUB WORK

4-H Club and FFA projects in hogs offer many opportunities for boys and girls to gain experience and knowledge in livestock management. There are three advantages of swine raising that appeal to the youngsters. Only a small outlay of capital is necessary. One can secure returns within a few months because of the rapid turnover. Hogs multiply quite rapidly and within a short space of time a sizeable herd can be built up.

The purchase of a bred gilt is the best way of getting started. In practically all areas of the Northeast there are hog breeders who have a surplus for sale. County agricultural agents and agricultural teachers will be glad to help the youngster find and select the gilts for his breeding herd.

Unless one has a preference, breed is not too important. It is always wise to select the breed that is the most popular in an area.

Since the offspring of the gilt will become the foundation of the new herd, careful selection should be made. Good breeding stock is necessary to obtain large litters of fast growing pigs of the desired type.

The gilt should be delivered to the farm well in advance of the farrowing date. This enables the 4-H member to get thorough training in the handling of gilts during the gestation period and through the critical farrowing time. On pages 12-26 of this book there is a detailed feeding and management discussion.

Showing

Show ring winning should not be the primary purpose of a pig club project, but it does add interest and the experience gained through competition is invaluable.

There are two divisions of showing for pig club members — breeding classes and fat classes. Individuals in either class should be selected well in advance of the fair and properly fitted and trained to handle.

Feeding For the Show Ring

All classes of livestock show up to better advantage if properly fitted. Breeding classes need not be in as high a degree of finish as the fat classes. The pigs must be well grown with a good degree of body finish at fair time. A high degree of body finish is necessary for the fat classes.

Training

Pigs shown in either the fat or breeding classes should be trained to handle easily with a cane or whip. They must be trained to walk, stop, turn and stand at will. Hogs being fitted for show should have ample exercise, preferably in small pasture lots. Older pigs must be driven to insure proper exercise being taken.

Fitting

The toes should be trimmed, if necessary, so that the hogs will stand erectly on their pasterns and the toes of each foot will be close together.

The pigs should be brushed daily with a stiff brush. If the weather is warm, they should be washed several times with soap and water to remove the scurf and scale. This also gives the skin and hair a much better appearance.

Before entrance into the show ring a light colored oil should be applied to the hair and skin, using just enough to make the hair glossy. Talcum powder may be used on white, belted or spotted pigs.

Ring Conduct

When the pig enters the show ring he should be looking his best and trained to respond to any movement that the owner wants. A good hog showman will keep his pig 20-25 feet away from the judge and always in as favorable a position as possible. A good showman will try to keep one eye on the hog and one eye on the judge so that the judge will never catch the showman or his pig out of position. The pig should be moved as little as possible, yet enough movement made so that the pig stays alert during the period he is in the show ring. Be ready to move at the judge's request. Also have any information such as age or other matter ready if asked by the judge. Accept the judge's decision and be a good loser or modest winner.

After the fairs the animals that will be used for breeding may be brought back and handled with the rest of the herd.

A TENTATIVE PLAN FOR 4-H CLUB OR F. F. A. PROJECT

FIRST YEAR

Purchase a good bred gilt to farrow in March or April.
Keep records of birth weight and weaning weight of baby pigs.
Wean pigs at eight weeks and breed the sow for fall farrowing.
Follow a good feeding program.
Select one or two of the best gilts for next year's farrowing, marketing the rest.
Keep accurate records of feed intake.
Market remainder of the litter.
Sell fall litter as weaning pigs, if facilities are not available for feeding during the winter.
Breed the original sow for spring farrowing, plus any gilts selected for the breeding herd.
Visit some successful hog raisers.

SECOND YEAR

Build portable farrowing houses if needed.
Plan a rotation system of pastures.
Keep records of offspring of each sow and select future breeding stock from the best litters only.
Cull out sows or gilts that produce small litters, are not good milkers or good mothers.
Select more gilts for breeding purposes if expansion can be handled.
Study ways of using labor-saving equipment.
Breed for fall litters and feed out litters if possible.
Summarize year's records and find weak spots that need correcting.

THIRD YEAR

Increase size of herd, if possible.
Study ways and means of saving labor.
Have a definite goal in your feeding program such as 200-pound pigs at five months of age.
Keep records and use these records to improve future operations.

CPSIA information can be obtained at www.ICGtesting.com
260245BV00001B/129/P

9 781446 536711